Earth's life support systems

How important are water and carbon to life on Earth?
Water and carbon support life on Earth and move between the land, oceans and atmosphere

Water is vital for the survival of organisms on Earth. It is needed for photosynthesis, respiration and transpiration and is the medium for chemical reactions in plants, animals and the soil. It is also an important economic resource.

Carbon is of great biological significance and all forms of life are based on large molecules of carbon atoms such as proteins and carbohydrates. It is also an important economic resource.

Water and carbon flow between stores — at a global scale both cycles are closed systems.

1 State three ways in which water regulates the thermal conditions of the Earth. *(3 marks)*

2 Identify the following carbon compounds: *(3 marks)*
 1 CH_4

 2 CO_2

 3 $CaCO_3$

3 State three ways in which carbon is used as an economic resource. *(3 marks)*

4 Explain why both water and carbon cycles can be regarded as closed systems at a global scale. *(2 marks)*

5 Explain how these two cycles can also exist as open systems. *(4 marks)*

The carbon and water cycles are systems with inputs, outputs and stores

The main stores in the carbon cycle are carbonate rocks, the oceans, sea floor sediments, fossil fuels, the atmosphere, plants and soils. The carbon cycle has two strands: a slow cycle and a fast cycle.

The water cycle is dominated by the oceans store, which contains 97% of all water. Other stores include ice, soil, groundwater, lakes/rivers, atmosphere and biosphere.

6 State the three main stores in the slow carbon cycle. *3 marks*

7 Using examples, describe three different states in which carbon exists. *3 marks*

8 Explain the role of weathering in the carbon cycle. *3 marks*

9 Using Figure 1, identify store X. *1 mark*

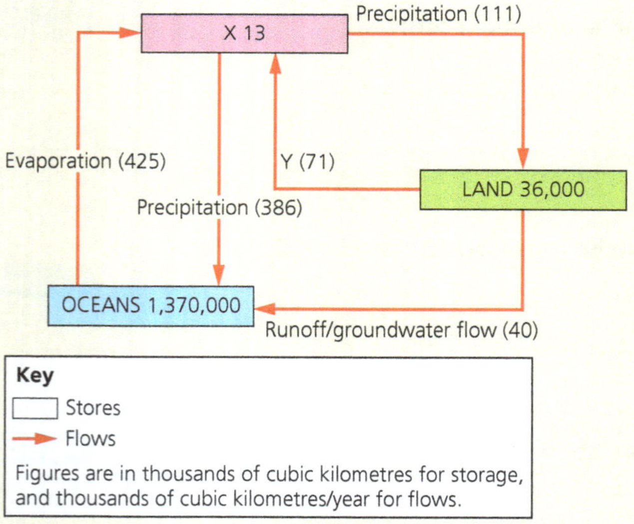

- X 13
- Precipitation (111)
- Evaporation (425)
- Y (71)
- Precipitation (386)
- LAND 36,000
- OCEANS 1,370,000
- Runoff/groundwater flow (40)

Key
- Stores
- Flows

Figures are in thousands of cubic kilometres for storage, and thousands of cubic kilometres/year for flows.

Figure 1

10 In what three states can water exist in stores? **3 marks**

11 What percentage of fresh water is stored in the cryosphere? **1 mark**

12 Explain why the amount of water stored in soils varies from place to place. **6 marks**

The carbon and water cycles have distinctive processes and pathways that operate within them

13 Using Figure 1 (from question 9), identify process Y. **1 mark**

14 Explain how water that evaporates from oceans can return to them. **4 marks**

15 State the water balance equation. **2 marks**

16 Explain why the annual input into a drainage basin may exceed the output. **3 marks**

17 Study Figure 2, which shows the global carbon cycle for the 1990s. The main annual fluxes are given in Giga tonnes of carbon (GtC)/year. Pre-industrial natural fluxes are shown in black and anthropogenic fluxes in red.

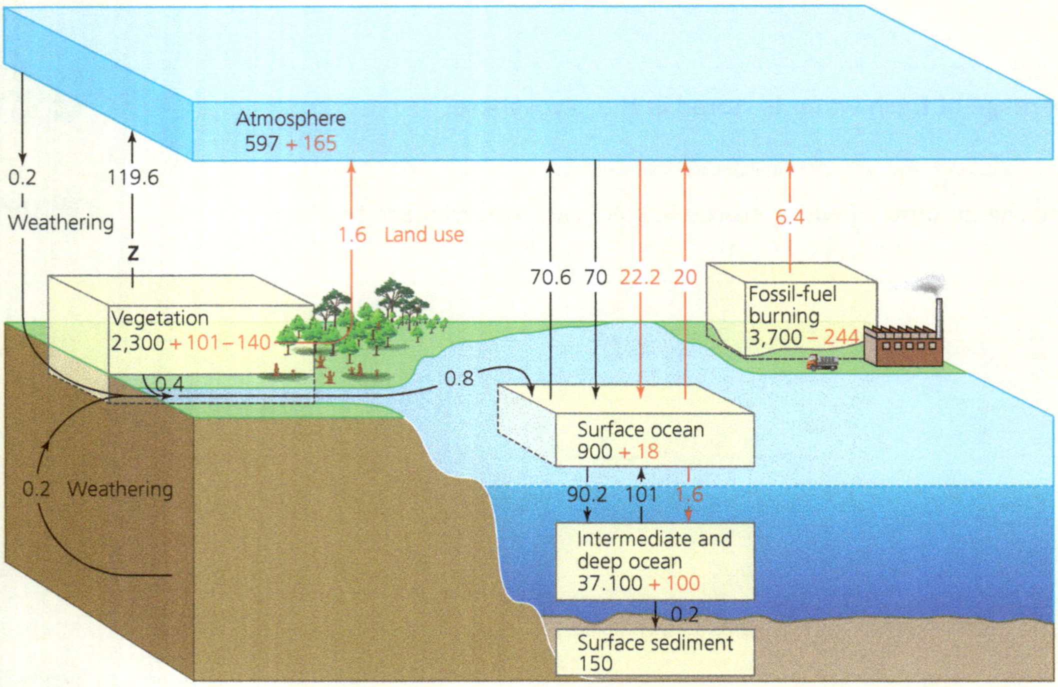

Figure 2

   a Identify process Z.   1 mark

   b Give the meaning of the following terms: 3 marks

Carbon sink	
Flux	
Sequestration	

18 Explain how carbon flows between the atmosphere and the ocean surface.   4 marks

How do the water and carbon cycles operate in contrasting locations?

It is possible to identify the physical and human factors that affect the water and carbon cycles in a tropical rainforest

You will study one tropical rainforest location for this key idea. You should know and understand:

- the rates of flows and distinctive stores of water and carbon
- how individual trees and the forest as a whole influence the cycles
- the physical factors affecting the water cycle
- the physical factors affecting the carbon cycle
- the changes to stores and flows in one drainage basin caused by natural and human factors
- the impact of human activity on carbon flows, soil and nutrient stores
- strategies to manage the tropical rainforest that have positive effects on the water and carbon cycles

Note: answers to some of these questions will depend on your chosen case study.

19 Study Table 1, which shows typical carbon stores in tropical rainforests.

	Carbon stored (tonnes of carbon/hectare)
Leaves of trees	5
Branches of trees	75
Trunks of trees	120
Small vegetation	10
Litter	10
Roots	30
Soil	100

Table 1

Calculate the percentage of total carbon stored:

a in living matter above ground **1 mark**

b in branches of trees **1 mark**

20 Explain how rock permeability influences the water cycle in your case study. **4 marks**

Earth's life support systems

7

21 Explain how the mineral composition of the rocks in your case study location influences the carbon cycle.

4 marks

22 Describe and explain how human factors have influenced the water cycle in a named drainage basin in the tropical rainforest.

6 marks

23 Explain the positive effects of two management strategies on the carbon cycle in your case study.

6 marks

It is possible to identify the physical and human factors that affect the water and carbon cycles in an Arctic tundra area

You will study one Arctic tundra location for this key idea. You should know and understand:

- the rates of flows and distinctive stores of water and carbon
- the physical factors affecting the water cycle
- the physical factors affecting the carbon cycle
- seasonal changes in the water and carbon cycles
- the impact of the developing oil and gas industry on the water and carbon cycles
- management strategies used to moderate the impacts of the oil and gas industry

Note: answers to some of these questions will depend on your chosen case study.

24 Explain how the temperature of the Arctic tundra influences terrestrial stores in the water cycle. *6 marks*

25 Explain how the biosphere stores in the carbon cycle of the Arctic tundra change seasonally. *6 marks*

26 Draw an annotated map of your case study area to describe and explain the location of the oil and gas industry. *6 marks*

27 Suggest reasons for two management strategies used to moderate the impacts of the oil and gas industry on the water and carbon cycles. `8 marks`

How much change occurs over time in the water and carbon cycles?

Human factors can disturb and enhance the natural processes and stores in the water and carbon cycles

28 What is meant by 'dynamic equilibrium'? `2 marks`

29 Outline one way in which the water cycle demonstrates a negative feedback loop. `2 marks`

30 Explain two ways in which land-use changes alter the water cycle. `4 marks`

31 What is an aquifer? — 2 marks

32 Explain why the height of the water table can vary over time. — 4 marks

33 Outline how human activities influence carbon sequestration. — 6 marks

Fast carbon cycle	Slow carbon cycle

34 Explain how an increase in global temperature can lead to both positive and negative feedback in the carbon cycle. — 6 marks

Positive feedback

Negative feedback

The pathways and processes which control the cycling of water and carbon vary over time

Study Figure 3, which shows atmospheric CO_2 concentration measured at the Mauna Loa observatory, Hawaii.

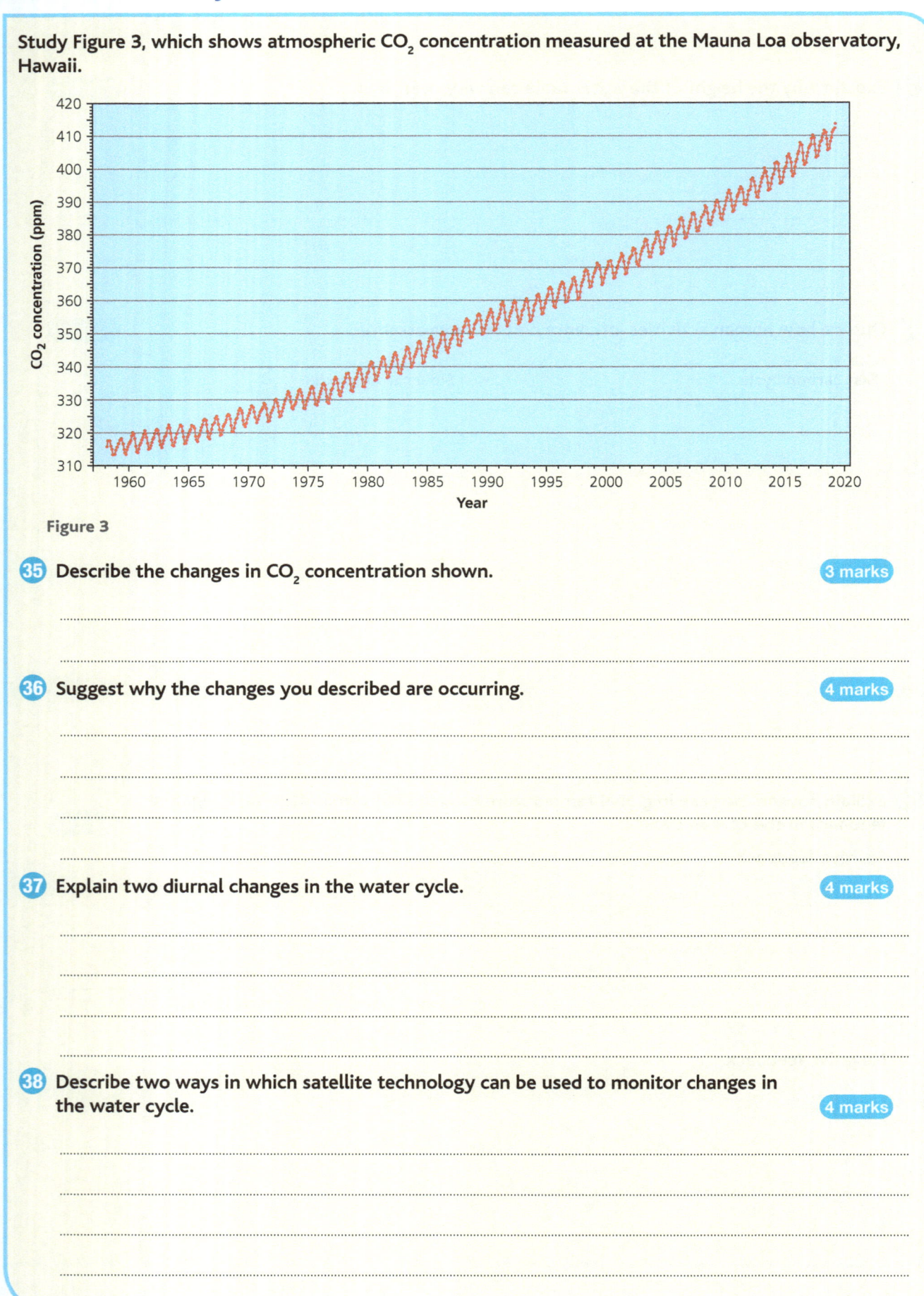

Figure 3

35 Describe the changes in CO_2 concentration shown. *3 marks*

36 Suggest why the changes you described are occurring. *4 marks*

37 Explain two diurnal changes in the water cycle. *4 marks*

38 Describe two ways in which satellite technology can be used to monitor changes in the water cycle. *4 marks*

39 How are glacial and interglacial periods related to changes in the carbon cycle? [2 marks]

40 Suggest one reason for this relationship. [4 marks]

To what extent are the water and carbon cycles linked?
The two cycles are linked and interdependent

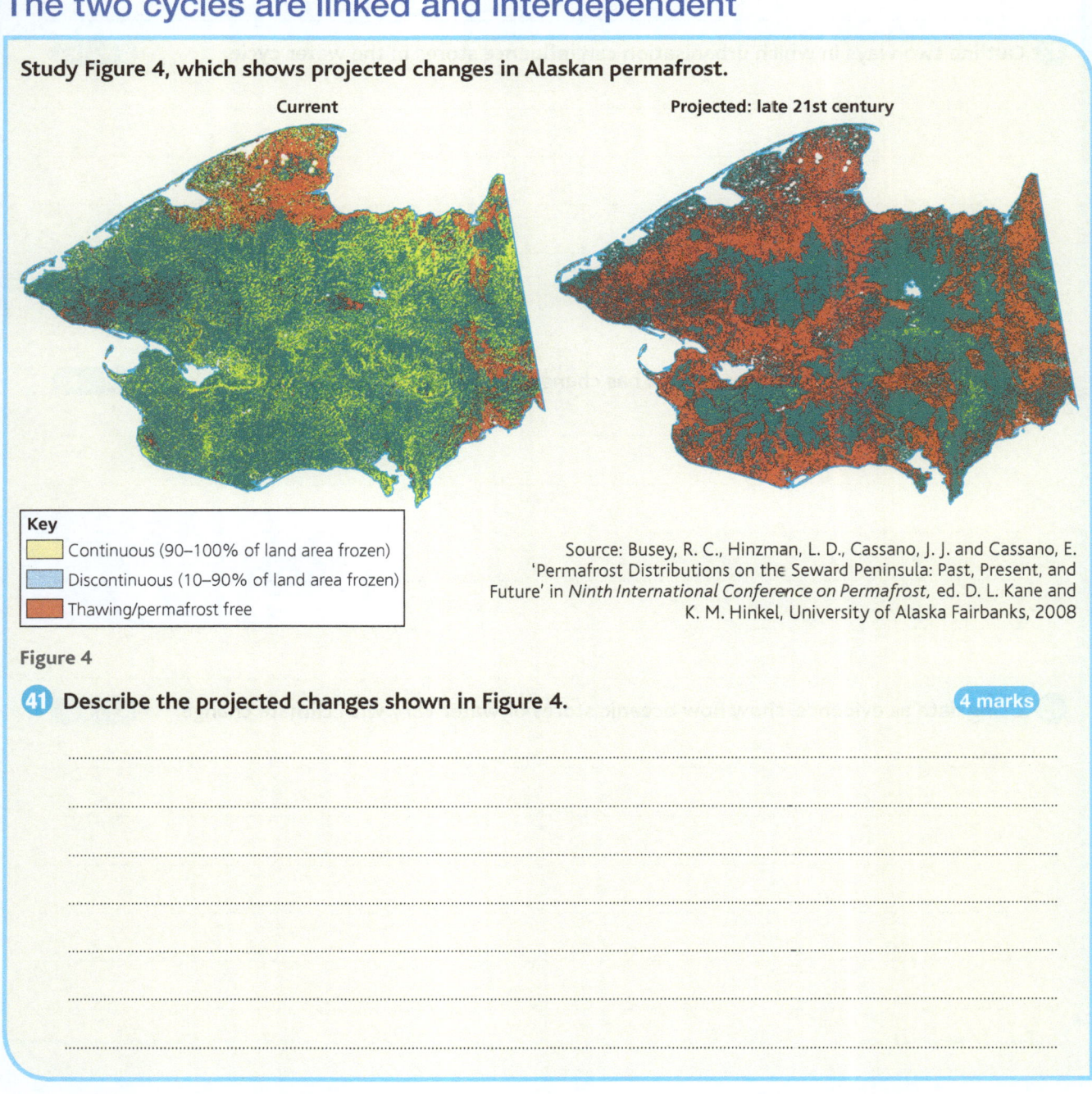

Figure 4

41 Describe the projected changes shown in Figure 4. [4 marks]

42 Explain how the changes you described link the carbon cycle and the water cycle. **6 marks**

43 Outline two ways in which urbanisation can influence stores in the water cycle. **4 marks**

44 Explain how human use of fossil fuels has changed stores in the carbon cycle. **4 marks**

45 Using data as evidence, show how oceanic stores of water vary with climate change. **4 marks**

The global implications of water and carbon management

Study Figure 5, which shows annual rates of afforestation in Ontario, Canada.

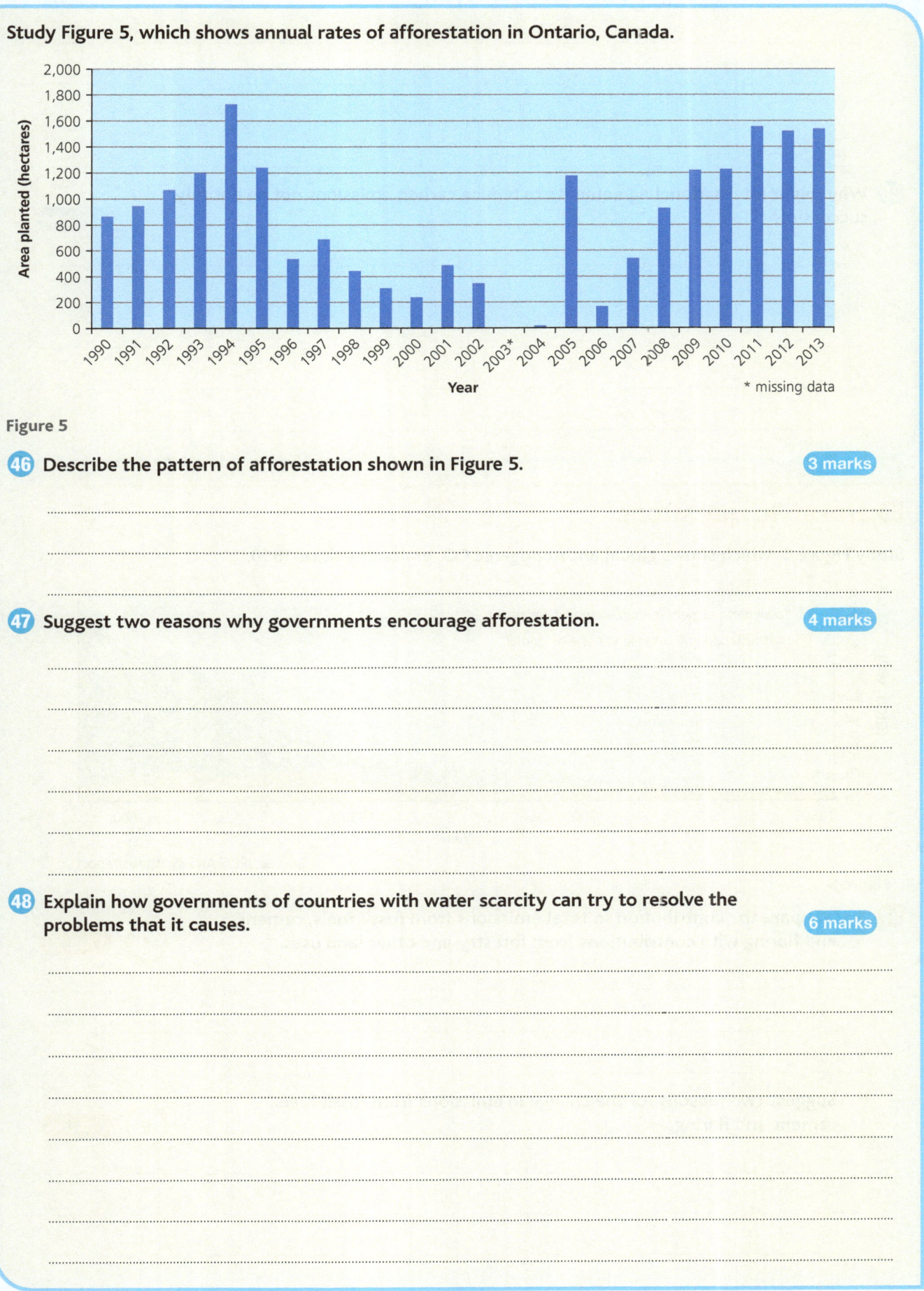

Figure 5

46 Describe the pattern of afforestation shown in Figure 5. **3 marks**

47 Suggest two reasons why governments encourage afforestation. **4 marks**

48 Explain how governments of countries with water scarcity can try to resolve the problems that it causes. **6 marks**

49 How can carbon trading influence the carbon cycle? 4 marks

50 Why might international agreements to reduce carbon emissions not be entirely successful? 6 marks

Exam-style questions

Study Figure 6, which shows global anthropogenic CO_2 emissions since 1850.

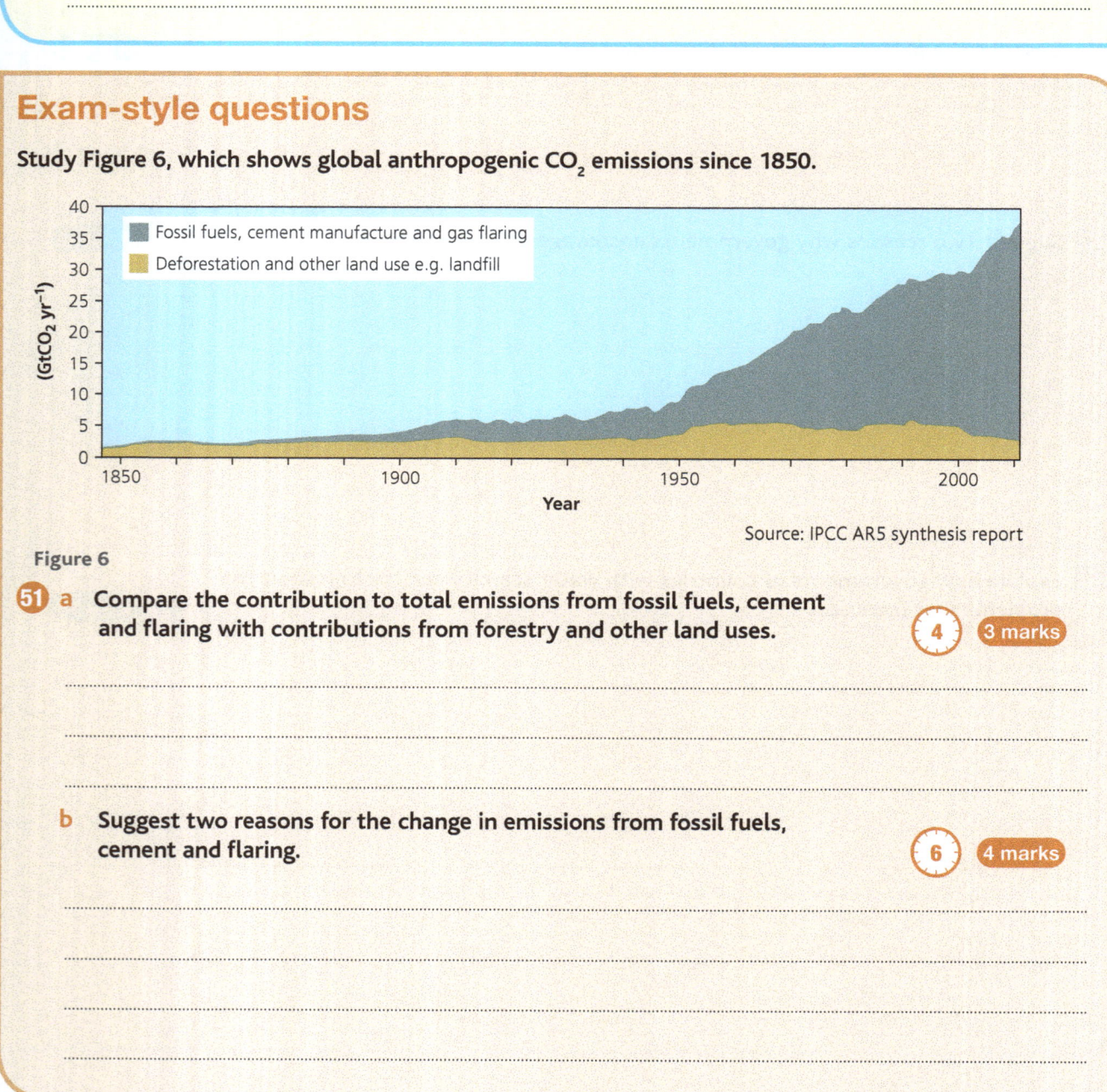

Figure 6

51 a Compare the contribution to total emissions from fossil fuels, cement and flaring with contributions from forestry and other land uses. 4 3 marks

b Suggest two reasons for the change in emissions from fossil fuels, cement and flaring. 6 4 marks

52 Examine the significance of seasonal changes in vegetation to the carbon cycle. (15) **10 marks**

53 'Improving forestry techniques is more effective in managing the global water cycle than water allocations'. How far do you agree? (20) **16 marks**

Global connections

Option A: Trade in the contemporary world

What are the contemporary patterns of international trade?

International trade involves flows of merchandise, services and capital, which vary spatially

Trade is an important part of the globalisation process and the subsequent integration of national economies. The international movement of goods (merchandise), services and capital has created a vast global trading system operating between countries and regions (such as Europe and North America). The global pattern of trade is uneven.

1. **Table 2 is a glossary of key terms associated with trade. Complete the table.** 4 marks

Table 2

Term	Definition
	Measured by average price of exports − average price of imports × 100
Economic multiplier	
	The difference between a country's inflows and outflows of money
Foreign Direct Investment (FDI)	

2. Study Figure 7, which shows the world share of exports of commercial services in 2018.

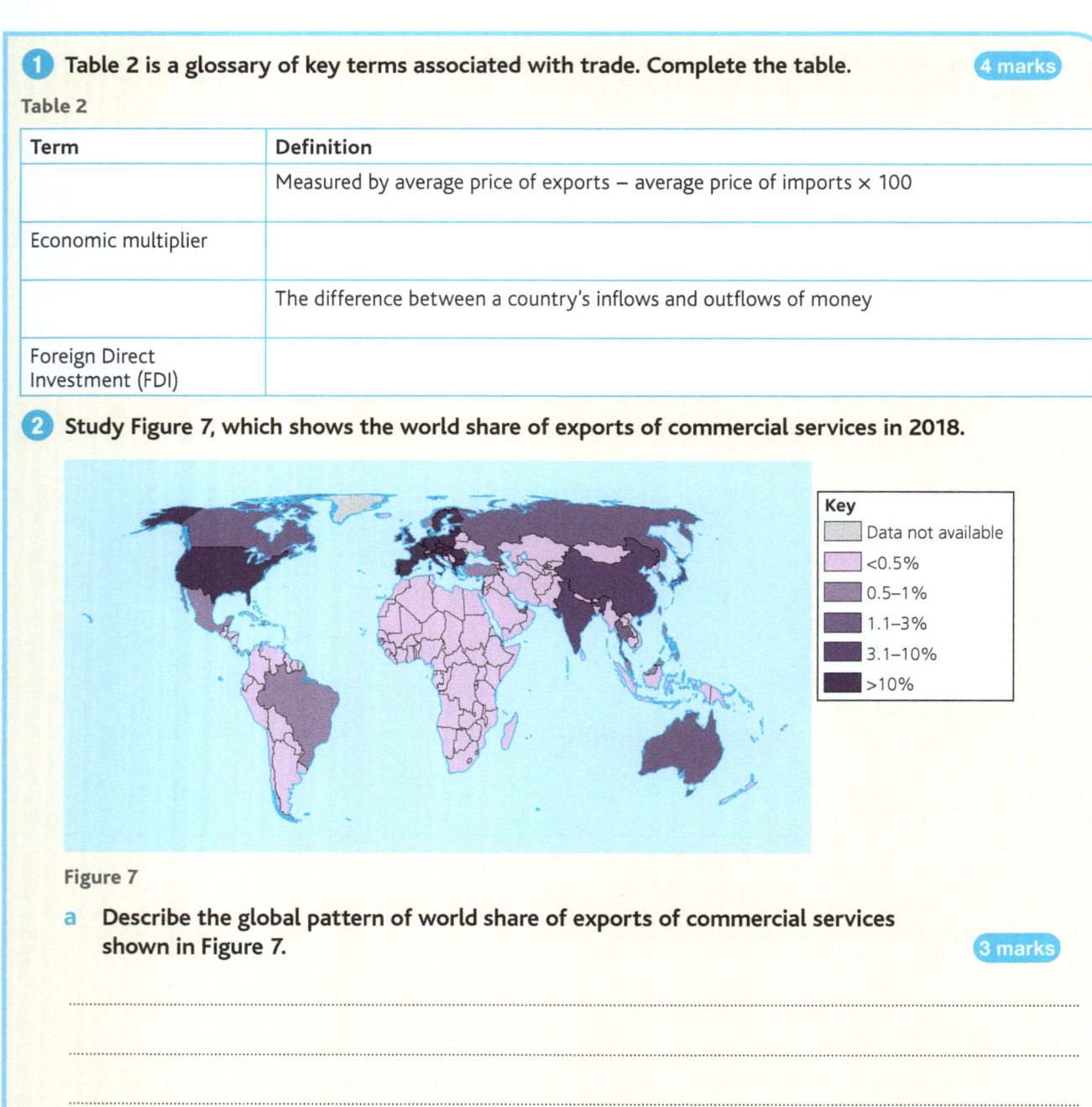

Figure 7

a Describe the global pattern of world share of exports of commercial services shown in Figure 7. 3 marks

....

....

....

b Suggest one reason for the low level of export of commercial services from Africa. **2 marks**

....

....

3 Referring to an example you have studied, explain the principle of comparative advantage. **3 marks**

....

....

....

....

4 State two advantages of being in a 'single market' economy. **2 marks**

....

....

5 Table 3 shows factors affecting contemporary patterns of international trade. Complete the table by adding two additional factors in each category. **8 marks**

Table 3

Economic	Political
Cost of production	Tariff/non-tariff barriers

Social	Environmental
Levels of education	Deep water ports

Current patterns of international trade are related to global patterns of socioeconomic development

There is a close relationship between trade and socioeconomic development. For many developing countries trade in regional and global markets offers benefits, however, trade also carries risks that can lead to inequalities, conflict and injustices. Many low-income developing countries (LIDCs) find access to global markets difficult. Therefore the transition from trade to growth and from growth to socioeconomic development is neither straightforward nor risk free.

6 Outline the nature of the relationship between the percentage share of world trade and the Human Development Index (HDI). **4 marks**

....

....

....

7 Explain why countries that are fully involved in international trade benefit from strong economic growth. **6 marks**

....
....
....
....
....
....
....
....

8 Table 4 shows disadvantages of international trade. Complete the table by explaining the relative disadvantages of international trade that have been identified. **10 marks**

Table 4

Loss of traditional skills	Labour exploitation	Uneven spatial growth within a country	Environmental damage	Protectionism and tariffs

Why has trade become increasingly complex?
Access to markets is influenced by a multitude of interrelated factors

International trade has become increasingly complex in the twenty-first century. Several factors interact to increase connectivity among suppliers and customers. These factors include: the effects of technology, transport and communications on global supply chains, the increasing global influence of multinational corporations (MNCs), the role of regional trading blocs; the growth of 'South–South' trade, the growth of services in the global economy and increasing labour mobility.

9 Distinguish between global supply chain, global value chain and outsourcing. **3 marks**

....
....
....

10 Outline the risks to companies who develop global supply chains. `6 marks`

11 How have developments in transport technology impacted on supply chains? `4 marks`

12 Identify three features common to MNCs. `3 marks`

13 Outline two advantages and two disadvantages of MNCs operations in host countries. `8 marks`

14 State two advantages for a country that is a member of a trading bloc such as the European Union (EU). `2 marks`

15 Account for the rise in South–South trade between developing countries in the twenty-first century. *(4 marks)*

....

....

....

....

16 Outline why growth of services is important in addressing poverty in LIDCs. *(4 marks)*

....

....

....

....

17 What is meant by the term New International Division of Labour (NIDL)? *(3 marks)*

....

....

....

18 Using Table 5, identify two positive factors and two restrictive factors influencing international labour mobility. *(4 marks)*

Table 5

Positive factors	Negative factors

There is interdependence between countries and their trading partners

Growing interdependence is a feature of nations that trade with each other. You will study one emerging and developing country (EDC) to illustrate:

- the direction and components of its current international trade partners
- changes in its international trade patterns over time
- its economic, political, social and environmental interdependence with trading partners
- the impacts of trade on the EDC, including economic development, political stability and social equality

19 With reference to an example of an EDC you have studied, illustrate how trading partners work together on environmental issues. **8 marks**

20 With reference to an EDC case study, explain the impact of trade on social equality. **6 marks**

What are the issues associated with unequal flows of international trade?

International trade creates opportunities and challenges which reflect unequal power relations between countries

In advanced countries (ACs) trade can lead to economic progress and political strength but can also present challenges such as trade disputes, negative environmental impacts and trade deficits. In LIDCs trade can present opportunities for development but also challenges such as internal inequality and political instability.

You will examine these through two case studies.

One AC to illustrate:
- its advantages for trade
- opportunities such as sustained economic growth
- challenges such as trade deficits

One LIDC to illustrate:
- its trade components
- why it has limited access to global markets
- opportunities such as diversification of economic activity
- challenges such as political instability

21 With reference to a named AC and LIDC that you have studied, contrast the environmental challenges created by international trade. Use the table below to set out your answer.

10 marks

AC case study:	LIDC case study:

22 Why is it important for LIDCs to become part of regional trading blocs?

6 marks

23 How do political factors limit access to global markets for a named LIDC you have studied?

8 marks

Exam-style questions

Table 6

Region	World merchandise exports (%)
North America	13.8
USA	9.0
Canada	2.4
Mexico	2.4
South and Central America and the Caribbean	3.4
Brazil	1.3
Chile	0.4
Europe	37.8
Germany	8.4
France	3.1
UK	2.6
Commonwealth and Independent States (CIS*)	3.0
Africa	2.4
South Africa	0.5
Middle East	5.6
Asia	34.0
China	13.2

*CIS: ten post-Soviet republics formed following the dissolution of the Soviet Union

24 a Study Table 6, which shows the percentage of world merchandise exports by region and selected countries, 2017.

  i State two factors that could account for the spatial variations shown in Table 6. **2 marks**

  ii Explain how the economic multiplier can be enhanced by international trade. **3 marks**

  iii The data shown in Table 6 are available every 10 years. Evaluate one method of data presentation that could be used to show changes in the percentage of world merchandise exports over time from 1947 to 2017. **4 marks**

b With reference to a case study of one EDC, examine the political nature of its interdependence with trading partners.

[10] 8 marks

25 Assess the opportunities and challenges resulting from international trade in a named LIDC you have studied.

[20] 16 marks

Option B: Global migration

What are the contemporary patterns of global migration?
Global migration involves dynamic flows of people between countries, regions and continents

Migration is population movement. The ongoing process of globalisation has meant that the shifts and flows of people over time and place have become increasingly complex. Migrant flows now reflect wide variations in scale, direction, motivation and composition. In terms of the duration of stay, timescales can be short or long term and distances travelled range from local cross-border to long-distance intercontinental movements. Migration has become an increasingly important issue for countries across the development spectrum.

1 Explain the link between globalisation and migration. *(3 marks)*

2 Define each of the following: *(3 marks)*
   a economic migrant

   b refugee

   c asylum seeker

3 How can international migration lead to population change within a country? *(3 marks)*

4 Give one current example of: *(3 marks)*
   a inter-regional migration

   b intra-regional migration

   c internal migration

Note: in this question 'region' refers to global regions, such as Europe, North America or Asia.

5 Explain the factors influencing one of your named examples from question 4. 6 marks

Current patterns of international migration are related to global patterns of socioeconomic development

A range of statistics and measures can be used to demonstrate the strong relationship between international migration and socioeconomic development. Such measures include the Human Development Index (HDI), receipt of migrant remittances and the extent to which economic remittances contribute to a country's gross domestic product (GDP). International migration also generates the flow of money, ideas and technology, which can promote stability, economic growth and development in countries of origin and destination but may also lead to conflicts, injustices and inequalities.

6 What is meant by the Human Development Index (HDI)? 2 marks

7 What are migrant remittances? 1 mark

8 Explain how the multiplier effect is stimulated by migrant remittances. 4 marks

9 Outline two ways in which international migration leads to conflict. 4 marks

10 Using Table 7, outline one advantage and one disadvantage of international migration for a host country and a country of origin in both the short and long term. **8 marks**

Table 7

	Advantage of migration	Disadvantage of migration
Short term	Host country	Host country
	Country of origin	Country of origin
Long term	Host country	Host country
	Country of origin	Country of origin

11 International migration increases the flow of ideas. How can this impact a country of origin? **4 marks**

....
....
....

12 a What is meant by the term diaspora? **1 mark**

....

b How can members of a diaspora promote socioeconomic development? **3 marks**

....
....
....

13 For any two of the following explain how they can be affected by immigration. **8 marks**

Population structure	Service provision, e.g. health and education services	Gross domestic product (GDP)	Unemployment

Why has migration become increasingly complex?

Global migration patterns are influenced by a multitude of interrelated factors

Global migration has become increasingly complex because of the interaction of social, economic, political and environmental factors.

14 a What is meant by the international migrant stock? *(1 mark)*

b Why have an increasing number of countries become interdependent through migrant flows? *(2 marks)*

15 Figure 8 shows female migrants as a percentage of international migrant stock in World Bank global regions. Describe the pattern shown. *(4 marks)*

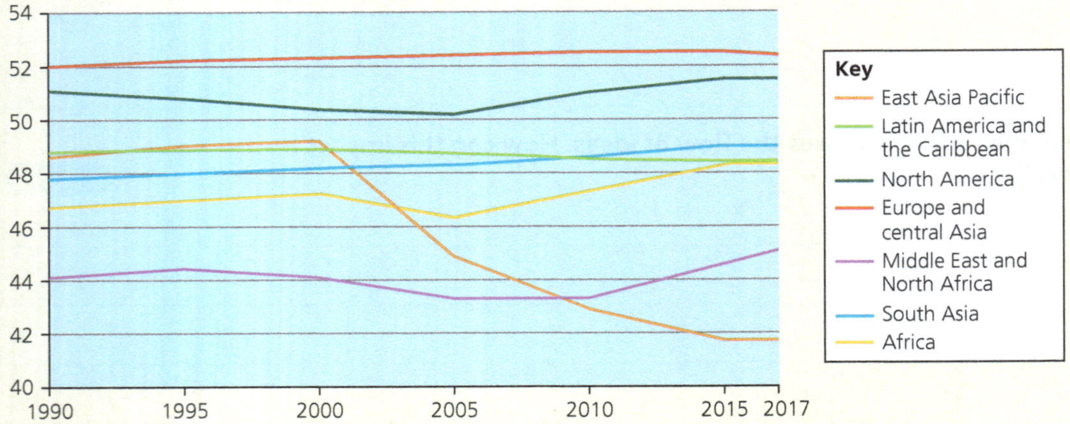

Key:
- East Asia Pacific
- Latin America and the Caribbean
- North America
- Europe and central Asia
- Middle East and North Africa
- South Asia
- Africa

Figure 8

16 Suggest three reasons for the pattern described in question 15. *(3 marks)*

17 Why has there been a significant rise in South–South migrant flows? `6 marks`

....
....
....
....
....
....

18 Outline the push and pull factors influencing one named South–South migration flow that you have studied. `4 marks`

....
....
....
....

19 State two reasons for the large number of refugees globally. `2 marks`

....
....

20 Table 8 shows the top three refugee-hosting countries in the world according to three different measures. Explain why measures relating to population size and GDP are valuable when analysing the impact of refugees on a host country. `3 marks`

Table 8

Total number of refugees	Refugees per 1,000 population	Refugees per US$1 million of GDP
Turkey (2.8 million)	Lebanon (173)	South Sudan (100)
Pakistan (1.6 million)	Jordan (89)	Chad (40)
Lebanon (1.0 million)	Turkey (35)	Lebanon (20)

Source: UNHCR

....
....
....
....

21 Suggest two reasons why 86% of refugees are hosted by low-income developing countries (LIDCs) and emerging and developing countries (EDCs). `4 marks`

....
....
....
....
....

22 **Compare and contrast the immigration policies of two named countries that you have studied. Use the table below to set out your answer.** *6 marks*

Country 1 name:	Country 2 name:

23 Suggest reasons why bilateral migration can cause political tensions between countries. *4 marks*

....

....

....

....

24 Referring to a named bilateral migration, outline one example of each of the following impacts on either country: *4 marks*

Named bilateral migration:

....

a economic impact

....

b social impact

....

c political impact

....

Corridors of migrant flows create interdependence between countries

Migration leads to increased interdependence between countries of origin and destination. On the positive side, countries may benefit from increased labour and the new skills, ideas and values that migrants bring. However, groups of people can also become marginalised by the effects of migration, leading to inequality and injustices.

You will study one EDC to illustrate:

- current patterns of immigration and emigration
- changes in immigration and emigration over time
- economic, political, social and environmental interdependence with countries connected to the EDC by migrant flows
- the impact of migration on the EDC's economic development, political stability and social equality

25 To what extent has migration between a named EDC and one of its bilateral migration partners created economic interdependence? *8 marks*

26 With reference to a named EDC, what effect has migration had on political stability? *6 marks*

What are the issues associated with unequal flows of global migration?

Global migration creates opportunities and challenges which reflect the unequal power relations between countries

Globally, some countries are able to drive international migration while others have a limited influence. These unequal power relations create a range of opportunities and challenges for both types of country. This must be explained through the study of one advanced country (AC) to show how it influences and drives change in the global migration system and one LIDC to show how it has limited influence over and restricted response to the global migration system. Your case studies should illustrate:

- patterns of emigration and immigration, migration policies and interdependence of countries linked to it by migration
- opportunities
- challenges

27 For a named AC you have studied:

  a Describe the patterns of emigration and immigration. *4 marks*

  b Outline the social factors that explain these patterns. *4 marks*

  c Examine the opportunities created as a result of international migration. *8 marks*

28 How do political factors affect migration policies in a named LIDC you have studied? *6 marks*

Exam-style questions

29 Study Figure 9, which shows the contribution of net international migration to population change in global regions.

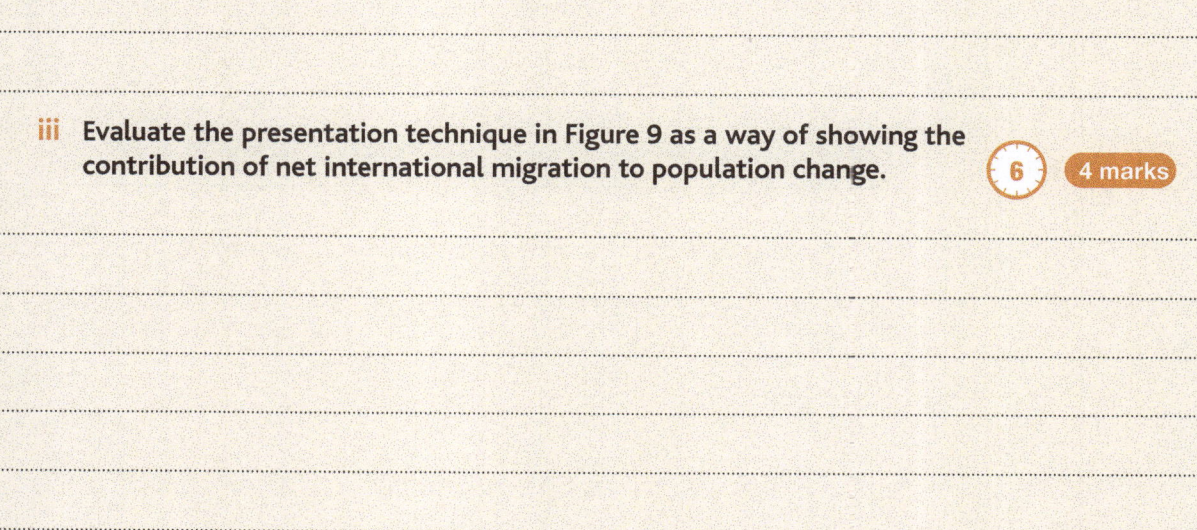

Figure 9

a i State two reasons why international migration has contributed to population growth in some regions while having a small or negative impact in others. **2 marks**

....

  ii Explain how the flow of ideas associated with migration can promote economic growth in regions with a high number of LIDCs such as Africa. **3 marks**

....
....

  iii Evaluate the presentation technique in Figure 9 as a way of showing the contribution of net international migration to population change. **4 marks**

....
....
....
....

b With reference to an EDC you have studied, examine the environmental nature of its interdependence with other countries as a result of migrant flows.

[8 marks]

Assess the opportunities and challenges created within a named LIDC as a result of international migration.

[16 marks]

Option C: Human rights

What is meant by human rights?
There is global variation in human rights norms

Humans have basic rights and freedoms that are in place to protect everyone, equally, at all times and in all places. However, there are wide geographical variations in human rights norms. Violations of human rights have occurred in many countries – rich and poor and at a variety of scales from individuals to large groups.

1 What is the UDHR? *(1 mark)*

2 What is meant by civil society? *(2 marks)*

3 Describe one way that globalisation has negatively affected the protection of human rights. *(2 marks)*

4 Complete Table 9 by outlining the meaning of each term in relation to human rights issues. *(9 marks)*

Table 9

Term	Meaning
Human rights norms	
Intervention	
Geopolitics	

5 State two forms of involvement by the United Nations when human rights are threatened. *(2 marks)*

Patterns of human rights violations are influenced by a range of factors

Article 3 of the UDHR states that 'Everyone has the right to life, liberty and security'. Global patterns of issues such as forced labour, maternal mortality and capital punishment vary because of contrasting perspectives on social, economic, political and environmental factors.

6 Give two examples of circumstances that would be classed as forced labour. *2 marks*

7 Explain three reasons for high levels of modern slavery in low-income developing countries. *6 marks*

8 Refer to Table 10, which shows maternal mortality ratio (MMR) and maternal deaths by selected global region, 2015.

Selected global regions	MMR (maternal deaths/ 100,000 live births)	Total number of maternal deaths due to childbirth
North Africa	70	3,100
Sub-Saharan Africa	546	201,000
Southeast Asia	110	13,000
West Asia (Middle East and east Africa)	91	4,700
Caucasus and central Asia (area south of Russia and west of China)	33	610
Latin America and the Caribbean	67	7,300
Oceania	187	500

Source: WHO

a With reference to Table 10, describe the pattern of MMR. *4 marks*

b Explain the existence of a high rate of MMR and maternal deaths among the countries of Sub-Saharan Africa.

4 marks

....
....
....
....
....
....

9 State two reasons why rates of capital punishment vary from one place to another.

2 marks

....
....

What are the variations in women's rights?
The geography of gender inequality is complex and contested

Gender inequality exists where individuals are treated differently based on their gender. Statistics show that although it is mainly women that suffer most, increasingly men and boys are also affected. Globally, wide variations in gender inequality can be seen but it is more prevalent in poorer countries where it remains a major obstacle to development.

You will study women's rights in a named country to illustrate:

- the gender inequality issues that are apparent in that country
- the consequences of gender inequality on society
- evidence of changing norms and strategies to address gender inequality issues

10 Figure 10 shows selected indicators for the global gender gap, 2008–17.

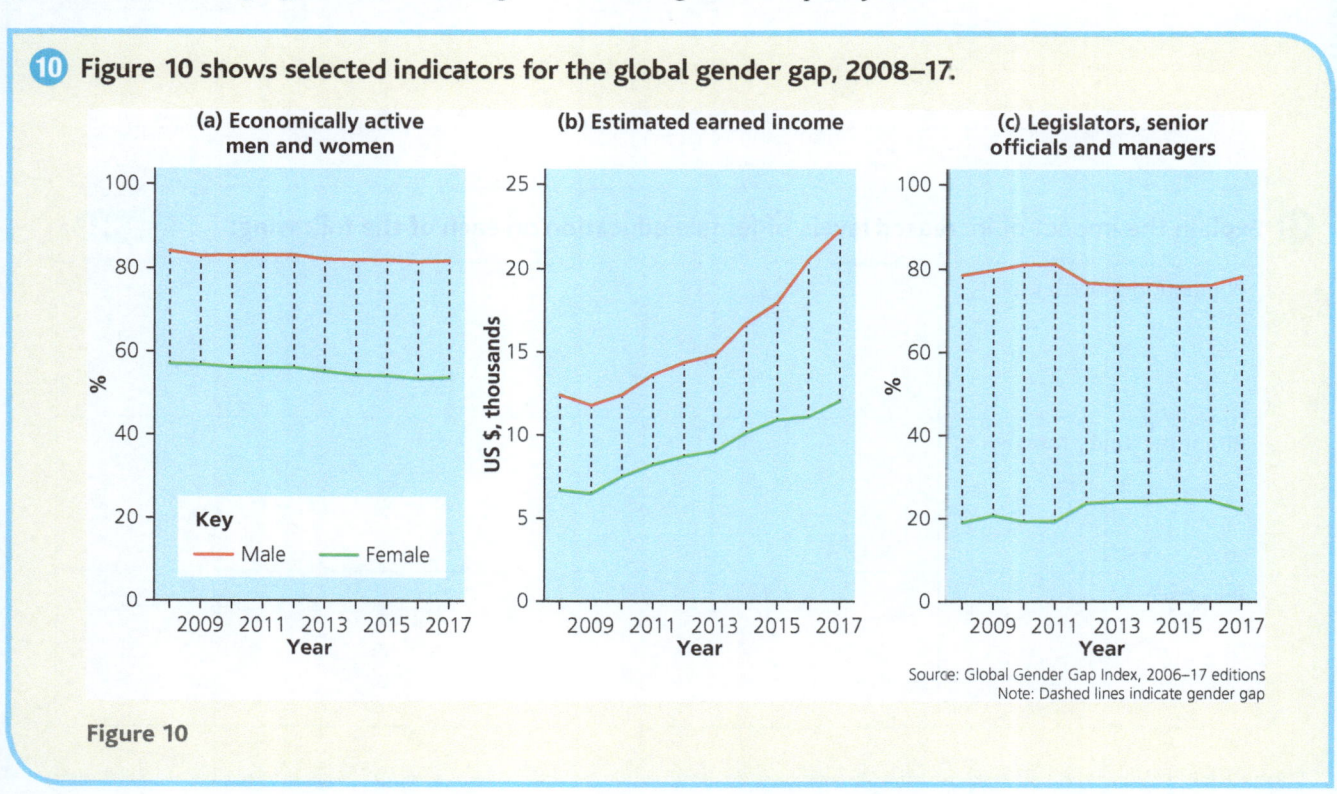

Figure 10

a   State three challenges women face as a result of gender inequality shown in Figure 10. **3 marks**

....
....
....

b   Name two indices that can be used to measure gender inequality. **2 marks**

....
....

11 Apart from individual governments, name one other type of organisation involved in resolving gender inequality issues. **1 mark**

....

12 State and explain the impact of two factors affecting female reproductive health in developing countries. **6 marks**

....
....
....
....
....
....
....

13 Why do many LIDCs show low levels of female participation in the labour market? **4 marks**

....
....
....
....

14 Explain the impact of increased levels of female education on each of the following: **6 marks**

Population growth rate	
Infant mortality rates	
Poverty	

15 For a named example you have studied, describe the consequences of gender inequality for society.

6 marks

16 a Outline the strategies used to address gender inequality issues in a named example you have studied.

6 marks

b State two obstacles limiting the implementation of these strategies.

2 marks

Obstacle 1:

Obstacle 2:

What are the strategies for global governance of human rights?

Human rights violations can be a cause and consequence of conflict

The violation of human rights can often be a factor in causing conflict. Once conflict starts, peoples' human rights are disregarded. Various measures can be used to restore human rights such as flows of people, money, technology and ideas.

17 State two examples of human rights violations that can cause conflict. `2 marks`

....

....

18 Suggest reasons why geopolitical intervention can be controversial. `2 marks`

....

....

....

19 Complete Table 11 by outlining the role of each of the four organisations named in the global governance of human rights. `6 marks`

Table 11

Institution	Role in the global governance of human rights
United Nations (UN)	
multinational corporations (MNCs)	
non-governmental organisations (NGOs)	
International Criminal Court (ICC)	

20 Suggest two reasons why some governments may not sign treaties established by supranational organisations such as the UN for the protection of human rights. `4 marks`

....

....

....

....

21 Explain the role of technology in geopolitical intervention. `4 marks`

....

....

....

....

22 How can interventions to reduce and prevent human rights violations be funded? `4 marks`

Global governance of human rights involves cooperation between organisations at scales from global to local, often in partnership

Institutions, treaties, laws and norms are established to protect human rights. A range of organisations cooperate to do this, from supranational (UN), regional (Association of Southeast Asian Nations – ASEAN) to local (NGOs).

These issues should be illustrated through your study of strategies for global governance of human rights in one named area of conflict, including:

- contribution and interaction of different organisations at a range of scales from global to local, including the UN, a national government and an NGO
- consequences of global governance of human rights for local communities

23 Describe three different motives behind military intervention to protect human rights in a named area of conflict that you have studied. `3 marks`

24 Referring to a named example, identify three disadvantages of military intervention. `3 marks`

25 Outline the benefits to local communities from the global governance of human rights in a named country you have studied. `6 marks`

To what extent has intervention in human rights contributed to development?

Global governance of human rights has consequences for citizens and places

Human rights violations hinder development whereas interventions can promote development. Global governance brings both positive and negative consequences for people and places in the short and long term. You will study one named LIDC to illustrate the consequence of global governance of human rights issues including:

- the human rights issue/issues
- the global governance strategy/strategies used
- opportunities for stability, growth and development
- challenges of inequality and injustice

For questions 26 to 28, refer to a case study of a named LIDC and use the tables below.

26 Outline the human rights issues. *(4 marks)*

Name of LIDC	Human rights issues

27 Identify and explain the strategies used to address human rights issues. *(6 marks)*

Strategy	Role in addressing human rights issues

28 Consider the short- and long-term effects resulting from global governance of human rights. *(8 marks)*

Short-term effects	Long-term effects

Exam-style questions

Figure 11

29 **a** Study Figure 11, which shows the Global Gender Gap Index, 2014.

  i State two factors that could account for the high level of gender inequality in some countries of the Middle East and parts of the African continent. **2 marks**

  ii Explain why female participation in education is very limited in many LIDCs. **3 marks**

  iii Evaluate the effectiveness of the data presentation technique used in Figure 11 to show the global gender gap. **4 marks**

b With reference to a case study of one area of conflict you have studied, examine how different organisations have contributed to global governance of human rights. [8 marks]

30 How does the global governance of human rights present challenges of inequality and injustice in a named LIDC you have studied? [16 marks]

Option D: Power and borders
What is meant by sovereignty and territorial integrity?
The world political map of sovereign nation-states is dynamic

Sovereign states dominate the global political system. Their boundaries can change, and new states form. Understanding the complexities of sovereignty and territorial integrity requires an appreciation of terms such as norms, intervention and geopolitics.

1 With reference to examples, distinguish between a state and a nation. **4 marks**

....
....
....
....

2 Define 'secession' and give one example. **2 marks**

....
....

3 Complete Table 12 by giving a definition of the key terms. **10 marks**

Table 12

Term	Definition
Norms	
International treaties	
Intervention	
Geopolitics	
Global governance	

4 What factors would lead to a state being described as resilient? **3 marks**

....
....
....

5 Explain the link between state fragility and development. 6 marks

6 Distinguish between sovereignty and territorial integrity. 4 marks

What are the contemporary challenges to sovereign state authority?

A multitude of factors pose challenges to sovereignty and territorial integrity

Nation-states are frequently challenged and contested. Reasons for such border disputes are many and varied. A complex range of economic, political, social and environmental factors combine to erode sovereignty and challenge territorial integrity.

You will examine these issues through the study of one country in which sovereignty has been challenged, including:

- causes and challenges to the government
- impacts on people and places

7 a Give two examples of contested borders. 2 marks

b Outline the factors that have contributed to one of the named examples in question 7a. 4 marks

8 What is meant by a TNC? *(2 marks)*

9 Assess the role of TNCs in the economic development of low-income developing countries (LIDCs). *(8 marks)*

10 How might organisations such as the United Nations (UN) and the Organisation for Economic Co-operation and Development (OECD) seek to control the negative impact of TNCs on sovereignty? *(2 marks)*

11 With reference to a named example of a trading bloc, outline the benefits and challenges to sovereignty for member states. *(6 marks)*

12 With reference to a named example, examine the challenges presented when a sovereign state includes more than one ethnic group. *(4 marks)*

13 Identify the main factors that have led to a challenge of the sovereignty and territorial integrity of a named country you have studied. **10 marks**

What is the role of global governance in conflict?
Global governance provides a framework to regulate the challenge of conflict

Challenges to sovereignty and/or territorial integrity often lead to conflict. Conflict is regulated by institutions such as the UN and NATO and the treaties, laws and norms that they establish in order to restore the global system of sovereign nation-states. Intervention by the international community is used in severe situations such as genocide, war and ethnic cleansing.

Human and financial resources are required for geopolitical intervention. These allow factors such as strategic thinking and surveillance technology to support global governance.

14 Identify three challenges to sovereignty and territorial integrity that lead to conflict. **3 marks**

15 Outline reasons why global governance is difficult to achieve in conflict situations. **4 marks**

16 Identify the main factors that organisations such as the UN must consider when planning intervention in conflict zones. `6 marks`

....
....
....
....
....
....
....

17 With reference to a named institution, how does it contribute to international security? `4 marks`

....
....
....
....

18 a List the four global commons. `4 marks`

....
....
....
....

b Why do they need global protection? `3 marks`

....
....
....

c State one way in which this protection is achieved. `2 marks`

....
....

Global governance involves cooperation between organisations at scales from global to local

Your study of the global governance in one area of conflict will illustrate:

- interventions and interactions of organisations at a range of scales, including the UN, a national government and a non-governmental organisation (NGO)
- consequences of global governance of the conflict for local communities

19 With reference to an area of conflict you have studied:

   a Outline the role of one NGO in global governance of the conflict. `5 marks`

   b To what extent did local communities benefit from global governance of the conflict? `6 marks`

How effective is global governance of sovereignty and territorial integrity?

Global governance of sovereignty and territorial integrity has consequences for citizens and places

Global governance has consequences that can be short and long term, and can bring about both intended benefits and also unintended negative impacts.

You will study the impact of global governance of sovereignty or territorial integrity in one LIDC to illustrate and explain:

- the sovereignty or territorial integrity issue/issues
- the global governance strategy/strategies used
- opportunities for stability, growth and development
- challenges of inequality and injustices

20 Examine three short-term effects of global governance where sovereignty has been threatened.

6 marks

21 How can global governance improve national systems upholding human rights?

4 marks

22 Describe the unintended effects of military intervention as part of global governance.

6 marks

23 State two short-term and two long-term benefits of global governance where territorial integrity has been threatened.

4 marks

Short-term benefits

1 ....

2 ....

Long-term benefits

1 ....

2 ....

24 Referring to a named LIDC you have studied:

a Explain the sovereignty/territorial integrity issues it faces. **4 marks**

b Outline the global governance strategies being used to address these issues. **4 marks**

c To what extent have the global government strategies presented further challenges of inequality and injustice? **8 marks**

Exam-style questions

Table 13

Ten most improved countries	Fragility index score	Ten most worsened countries	Fragility index score
Haiti (−3.3)	102.0	Qatar (+4.1)	48.1
Iraq (−3.2)	102.2	Spain (+3.5)	41.4
Nepal (−3.1)	87.9	Venezuela (+3.3)	86.2
Ecuador (−3.1)	74.2	USA (+2.1)	37.7
Japan (−2.9)	34.5	Yemen (+1.6)	112.7
Mexico (−2.8)	71.5	Turkey (+1.4)	82.2
Senegal (−2.7)	79.6	Togo (+1.3)	85.2
Seychelles (−2.6)	56.8	Bangladesh (+1.2)	90.3
Luxembourg (−2.6)	20.8	Philippines (+1.1)	85.5
Kuwait (−2.6)	55.9	UK (+1.1)	34.3

Fragility index is in the range 10–120 — 10 = sustainable, 120 = alert